LEAN SIGMA MASTERY GUIDE

The ultimte Guide On How To Change Your mindset On Continuous Improvement, Team Input, Scope Flexibility, Delivering Essential Quality Products In Project Management and more

GEREMY WILSON

© **Copyright 2021 - All rights reserved.**

The content contained within this book may not be reproduced, duplicated or transmitted without direct written permission from the author or the publisher.

Under no circumstances will any blame or legal responsibility be held against the publisher, or author, for any damages, reparation, or monetary loss due to the information contained within this book. Either directly or indirectly.

Legal Notice:

This book is copyright protected. This book is only for personal use. You cannot amend, distribute, sell, use, quote or paraphrase any part, or the content within this book, without the consent of the author or publisher.

Disclaimer Notice:

Please note the information contained within this document is for educational and entertainment purposes only. All effort has been executed to present accurate, up to date, and reliable, complete information. No warranties of any kind are declared or implied. Readers acknowledge that the author is not engaging in the rendering of legal, financial, medical or professional advice. The content within this book has been derived from various sources. Please consult a licensed professional before attempting any techniques outlined in this book.

By reading this document, the reader agrees that under no circumstances is the author responsible for any losses, direct or indirect, which are incurred as a result of the use of information contained within this document, including, but not limited to, errors, omissions, or inaccuracies.

Table of Contents

INTRODUCTION 7
   Why is Kaizen Effective? 25
   Kaizen Improvements: Plan-Do-Check-Act (PDCA) Process 26
   Identifying Problems 28
   Setting New Standards 29
   The System for Welcoming Suggestions 29
   Process Focused Mindset 32
   Innovation vs. Kaizen 33
   Application of Kaizen in the Toyota Production System 33
Chapter 1: What is Lean Analytics, A General Overview and a Little Bit of History about It 45
Chapter 2: Lean Thinking 62
CONCLUSION 77

INTRODUCTION

Lean Analytics

Lean Analytics is part of the methodology for a lean startup, and it consists of three elements: building, measuring, and learning. These elements are going to form up a Lean Analytics Cycle of product development, which will quickly build up to an MVP, or Minimum Viable Product. When done properly, it can help you to make smart decisions provided you use the measurements that are accurate with Lean Analytics.

Remember, Lean Analytics is just a part of the Lean startup methodology. Thus, it will only cover a part of the entire Lean methodology. Specifically, Lean Analytics will focus on the part of the cycle that discusses measurements and learning.

It is never a good idea to just jump in and hope that things turn out well for you. The Lean methodology is all about experimenting and finding out exactly what your customers want. This helps you to feel confident that you are providing your customers with a product you know they want. Lean Analytics is an important step to ensuring that you get all the information you need to make these important decisions.

Before your company decides to apply this methodology, you must clearly know what you need to track, why you are tracking it, and the techniques you are using to track it.

Focus on the fundamentals

There are several principles of Lean that you will need to focus on when you work with Lean Analytics. These include:

- A strive for perfection
- A system for pull through
- Maintain the flow of the business
- Work to improve the value stream by purging all types of waste
- Respect and engage the people or the customers
- Focus on delivering as much value to the customer as effectively as possible

Waste and the Lean System

One of the most significant things that you will be addressing with Lean Analytics, or with any of the other parts of the Lean methodology, is waste. Waste is going to cost a company time and money and often frustrates the customer in the process. Whether it is because of product construction, defects, overproduction, or poor customer service, it ends up harming the company's bottom line.

There are several different types of waste that you will address when working with the Lean system. The most common types that you will encounter with your Lean Analytics include:

- **Logistics:** Take a look at the way the business handles the transportation of the service or product. You can see if there is an unnecessary movement of information, materials, or parts in the different sections of the process. These unnecessary steps and movements can end up costing your business a lot of money, especially if they are repeated on a regular basis. This will help you see if more efficient methods exist.

- **Waiting:** Are facilities, systems, parts, or people idle? Do people spend much of their time without tasks despite the availability of work or do facilities stay empty? Inefficient conditions can cost the business a lot of money while each part waits for the work cycle to finish. You want to make sure that your workers are taking the optimal steps to get the work done, without having to waste time and energy.

- **Overproduction:** Here, you'll need to take a look at customer demand and determine whether production matches this demand or is in excess. Check if the creation of the product is faster or in a larger quantity than the customer's demand. Any time that you make more products than the customer needs, you are going to run into trouble with spending too much on those products. As a business, you need to learn what your customer wants and needs, so you make just the amount that you can sell.

- **Defects:** Determine the parts of the process that may result in an unacceptable product or service for the customer. If defects do exist, decide whether you should refocus to ensure that money is not lost.

- **Inventory:** Take a look at the entire inventory, including both finished and unfinished products. Check for any pending work, raw materials, or finished goods that are not being used and do not have value to them.

- **Movement:** You can also look to see if there is any wasted movement, particularly with goods, equipment, people, and materials. If there is, can you find ways to reduce this waste to help save money?

- **Extra processing:** Look into any existing extra work, and how much is performed beyond the standard that is required by the customer. Extra processing can ensure that you are not putting in any more time and money than what is needed.

How Lean can help you define and then improve a value stream

Any time that you look at the value stream, you will see all the information, people, materials, and activities that need to flow and cooperate to provide value to your customers. You need these to come together well so that the customer gets the value they expect, and at the time and way, they want it. Identifying the value stream will be possible by using a value stream map.

You can improve your value stream with the Plan-Do-Check-Act process.

Another method of creating this environment is the 5S+ (Five S plus): sort, straighten, scrub, systematize, and standardize. Afterward, ensure that any unsafe conditions along the way are eliminated.

The reason that you will want to do the sorting and cleaning is to make it easier to detect any waste. When everything is a mess, and everyone is having trouble figuring out what goes where, sorting and cleaning can address waste quite fast. There will also be times when you deem something as waste and then find out that it is actually important.

When everything is straightened out, you can make more sense of the processes in front of you. Afterward, you can take some time to look deeper into the system and eliminate anything that might be considered as waste or unsafe, and spend your time and money on parts of the process that actually provide value for your customer.

The Lean Analytics Stages Each Company Needs to Follow

To be successful with Lean Analytics, you'll need to follow several different stages. You won't be able to move on to the next stage if you do not complete the preceding step. There are five in particular that you will need to focus on to get work done with this section of the Lean support methodology. The five stages are:

- **Stage 1:** The initial stage is where you will concentrate on finding the problem for which people are searching for a solution. A business that focuses on business to business selling is going to find this stage critical. When you address this problem, then you can move on to the next stage.

- **Stage 2:** For this stage, you are going to create an MVP product that can be used by early adopter customers. This stage is where you are aiming for user retention and engagement, and you can spend some time learning how this will happen when people start to use the product. You can also learn this information based on how the customer uses your site and how long they stay. You'll take some time at this stage because you will need to experiment and also may need to go through and choose from a few different products before you get the one that is right for you. Once you have this information, you can move on.

- **Stage 3:** Once you find out how the early adopter customers are going to respond to a product or service, it is time to find the most cost-efficient way to reach more customers. Once you have a plan ready to get those customers, and then more of them start purchasing the product, then you can move to the next stage. You would not want to go with a product that may be popular but costs a ton of money, which will cut into your revenues and can make it difficult to keep growing in the future.

- **Stage 4:** You are now going to spend some time on economics and focusing on how much revenue you are making. You want to be able to optimize the revenue, so you need to calculate out the LTV:CAC ratio. LTV is the revenue that you expect to get from the customer, and the CAC is the cost that you incurred to acquire that customer. You can find this ratio by dividing your LTV by the CAC. Your margins are doing well if you get an LTV that is three times higher than the CAC. The higher the margins you get, the better because that means you are going to earn more in profits from the endeavor.

- **Stage 5:** In the final stage, you will then take actions that are necessary to grow the business. You can continue with your current plan if you are making a high enough margin from the previous steps, or you may need to make some changes to ensure that you can earn enough revenue to keep the business growing. You can also spend time making plans on where you would like to concentrate on in the future to increase the growth of your business and help it expand. The main goal for your business is to keep growing and increase revenue. This step helps you to reevaluate what you have in your current plan and decide if it is working for you or if you need to go with a different option.

Scrum is decisively a development of Agile Management. Scrum strategy depends on a lot of characterized practices and jobs that must be included during the product advancement process. It is an adaptable approach that rewards the use of the 12 spry standards in a setting concurred by all the colleagues of the item.

Scrum is executed in brief hinders that are short and intermittent, called Sprints, which generally run from 2 to about a month, which is the term for criticism and reflection. Each Sprint is an element in itself, that is, it gives a total result, a variety of the last item that must have the option to be conveyed to the customer with the least conceivable exertion when mentioned.

The procedure has as a beginning stage, a rundown of targets/necessities that make up the task plan. It is the customer of the task that organizes these goals considering an equalization of the worth and the expense thereof that is the manner by which the emphases and resulting conveyances are resolved.

From one perspective the market requests quality, quick conveyance at lower costs, for which an organization must be extremely deft and adaptable in the advancement of items, to accomplish short improvement cycles that can fulfill the need of clients without undermining the nature of the outcome. It is an extremely simple technique to actualize and well known for the snappy outcomes it gets.

Scrum technique is utilized for the most part for programming advancement, however different divisions are additionally exploiting its advantages by actualizing this procedure in their hierarchical models, for example, deals, showcasing, and HR groups and so on.

Scrum Development

In Scrum, the group centers around building quality programming. The proprietor of a Scrum venture center around characterizing what are the attributes that the item should need to manufacture (what to fabricate, what not, and in what request) and to defeat any deterrent that could ruin the undertaking of the improvement group.

The Scrum group comprises of the accompanying jobs:

Scrum ace: The individual who drives the group managing them to agree to the standards and procedures of the system. Scrum ace deals with the decrease of obstructions of the undertaking and works with the Product Owner to expand the ROI. The Scrum Master is responsible for staying up with the latest, giving instructing, tutoring and preparing to the groups on the off chance that it needs it.

Item proprietor (PO): Is the delegate of the partners and clients who utilize the product. They center around the business part and is liable for the ROI of the task. They Translate the vision of the venture to the group, approve the advantages in stories to be fused into the Product Backlog and organize them all the time.

Group: A gathering of experts with the fundamental specialized information who build up the task together doing the narratives they focus on toward the beginning of each dash.

Applications of Scrum

Scrum has been used worldwide extensively and applied across various use cases including but not limited to: research and identify markets, technologies, and product capabilities; develop and release products and enhancements as frequently as many times per day; maintain and sustain products, systems, and other operational environments. Further, Scrum has been used to develop software (embedded and otherwise), hardware, networks of interacting functions, autonomous vehicles, schools, government, non-profit organizations, marketing, operations, and almost everything we use in our daily lives.

Fundamental Scrum Trade-offs

There are four fundamental trade-offs defined by the Agile Manifesto that Scrum Framework implements:

1. Individuals and interactions OVER processes and tools

2. Working software OVER comprehensive documentation

3. Customer collaboration OVER contract negotiation

4. Responding to change OVER following a plan

Furthermore, there are three main focus areas that the Scrum Framework implements:

1. <u>Focus on value:</u> Everything that is done with the *Agile Mindset* focuses on the value it creates. If there is value, then do it. If there is no value, then don't do it.

2. <u>Focus on collaboration:</u> Scrum Framework focuses on teaming the people with the right skills and the right mindset for creative collaboration by providing the right cultural environment to enable and amplify strong collaboration.

3. <u>Focus on adaptability:</u> Scrum Framework deals with the fact that requirements do change quickly and frequently. Therefore, teams re/de-prioritize existing work when it is realized that it is not valuable. For that reason, the *Agile Mindset* gives special emphasis on adaptability.

The Background of Scrum

You can apply these principles to your life to create an immediate sense of accomplishment. Although you may accomplish some of the goals you set, these goals may not be aligned with your values. Plus, if you don't recognize accomplishments regularly, you'll likely lose enthusiasm or get distracted easily.

Using this method, your definition of success will shift from "what can I accomplish this year?" to "what can I do to move closer to my goals today?" Every day becomes a successful day, and at the end of four weeks, just 28 days, you have something tangible to call success. It's inspiring and self-reinforcing. You become unstoppable.

You also become more flexible. If life throws you a curveball, it's easy to adjust because you are nimble and focused on your values rather than some obscure, immovable object in the future.

Kanban is a visual workflow management method for managing work in an effective manner. It is a popular lean framework used to eliminate waste and organize work.

Kanban visualizes both the workflow process and work items flowing through the workflow. The visual representation of work items on a Kanban board allows team members to know the current state of every work item at any given time.

Kanban originally was designed for the manufacturing industry to reduce the idle time in a production process, but over time, it has become an efficient way for delivering products of any industry.

In a manufacturing process, Kanban covers the end to end process flow – from supplier to the end consumer. This end to end process visualization helps to avoid disruption or delays at any stage of the manufacturing process and minimizes overstocking of goods in the process.

In a traditional project management approach, developers are often overworked to complete work per the aligned plan. On the contrary, the core purpose of Kanban is to ensure a continuous delivery without overburdening the development team.

Kanban visualizes the flow of work items across different states thereby identifying potential bottlenecks in the process flow. Thus, with Kanban approach, the development team can proactively fix the potential impediments so work items can continually flow through the states at an optimal speed.

Kanban is a Japanese word that means "visual card" or "signboard". In the automotive sector, Kanban cards play an important role in tracking production within a factory. They signal the need to move materials within a factory. These cards also tell the supplier when to deliver a new shipment, thereby bringing visibility of the manufacturing process to suppliers. Kanban cards trigger a replenishment of the product or inventory based on the demand or the need of the specific product or inventory. Kanban serves as a scheduling system that manages what, when, and how much to produce.

Kanban is based on just-in-time (JIT) manufacturing principles. Just-in-time manufacturing, or JIT, is a management philosophy to eliminate manufacturing waste of finished products, half-finished products, parts, and supplies by producing only the right amount of parts at the right place at the right time. This concept minimizes the need to store inventory, which adds to the cost of the product.

JIT was developed by Taiichi Ohno of Toyota, who is now called as the father of JIT. JIT system is basically a "pull system", which means that the production in one system is dependent on the demand from the next system. Thus, if the next system does not need a specific part or product, it will not be manufactured. With this approach, the production of a product, or a part is completely dependent on the end-user needs. This concept is also known as on-demand production.

Kanban facilitates the execution of a just-in-time (JIT) system by attaching visual cards or Kanban cards to every stage of a production or a process flow. Each card depicts the quantity and amount of inventory that is required to be manufactured at that stage in the production system.

Apart from JIT, Kanban is also based on lean methodology. This methodology aims to maximize customer value while eliminating waste. Anything that does not add value to the customer, such as excessive inventory, inefficient meetings, unnecessary documentation, etc. is considered a waste. Lean focuses on eliminating such waste to continually improve the organizational system.

Kanban enables continuous improvement and lean concepts through a visual workflow management system and a disciplined approach to work. Kanban boards are visible to all stakeholders and represent a real-time state of work within a team or a portfolio.

Kanban cards on the board represent the actual work being completed, the assignee of the work item, and the rough estimate to complete the same in terms of hours.

To summarize, Kanban is a visual workflow management tool or a scheduling system based on the principles of lean and just-in-time production.

Origins of Kanban

In the early 1940s, Taiichi Ohno, a Japanese industrial engineer, created a simple system to manage work and control inventory at every stage of production at Toyota. This production system was known as the Kanban system. Kanban was developed to achieve an efficient just-in-time production system while reducing cost-intensive waste.

Toyota line workers used Kanban cards to send signals from one manufacturing unit to another. The signals denoted the need to move inventory or parts or a need to replenish materials from an external supplier into the factory. Since Kanban is based on a "pull system", each card denoted a depletion of a part of inventory and signaled replenishment of the same to match inventory with demand. These cards served as visual cues to effectively manage their production thereby increasing quality and reducing production cost.

Kaizen is a union of the two Japanese terms 'Kai' (continuous) and 'Zen' (improvement). Hence, the word Kaizen is translated as continuous improvement. Other translations refer to 'Kai' as change while 'Zen' refers to good. So, Kaizen can also mean change for the good or for the better.

Regardless of the variations in the translation, Kaizen is one of the most popular words in Japan. You can read it in the newspapers, hear it over the radio, watch in on the television, and encounter it on social media.

Every day, the Japanese are bombarded with statements about the Kaizen of everything. In the world of business, the principle of Kaizen has been deeply oriented in the core of workers and managers that they don't even need to think about Kaizen.

The primary difference between how you understand change in Japan and how it is perceived in the United States can be found in the philosophy of Kaizen. This philosophy is inherent to many Japanese that they don't even know that they have been doing it. This is the reason why many businesses in Japan are always changing.

In the philosophy of Kaizen, not a day should pass by without some form of change being made in the business. After the Second World War, many Japanese companies literally had to start from the ashes. Each day brought new problems to solve for both workers and managers, and each day offers a new set of challenges. The mere fact of staying in business needs a continuous cycle of progress, and Kaizen has helped the Japanese society to take on any challenge.

Kaizen can help you improve and transform your life one step at a time. But in order to do that, you should first understand the origin and application of this philosophy.

Why is Kaizen Effective?
Kaizen is an effective management system because it helps the organization to reduce muds (waste) and discard processes that are murk (laborious). As a practice for lean business, Kaizen is an effective process once all persons

involved are looking for business areas to improve and give recommendations based on their experience and observations. Basically, these recommendations are for minor changes, which could gradually improve the business for good (Kaizen).

From the start, it should be clear that all recommendations will be accepted and that there will be no repercussions for participating and for voicing out their observations. Rather, employees are praised and given rewards for the changes that are proven to improve the workplace. For instance, Toyota encourages its staff to make recommendations by offering small perks for every change made.

Another form of intangible reward is when the staff sees their recommendations implemented to the business. People will become more confident and will increase their sense of belongingness. Regardless of their role in the company, they become leaders who are constantly looking for areas, which could be improved.

Kaizen Improvements: Plan-Do-Check-Act (PDCA) Process

By using the PDCA process in implementing changes will make certain that there will be constant cycle in place to keep track of the changes and keep on improving them.

PLAN: Identify a problem and find possible solutions

DO: Act on the best solution

CHECK: Assess the results to know if the solution is effective

ACT: If the solution is effective, standardize it and implement it for the whole organization. If the solution is not effective, think it over again and plan again

Kaizen offers a basic philosophy: assess how you can improve things, act on them, and then continue the improvement process constantly. The broad nature of Kaizen means that it is highly relevant to other tools for lean management that you can use in the continuous improvement process.

The following lean business tools are categorized under the Kaizen clout:

5S - Use the 5S system for constant improvement by attaining organization and cleanliness for the whole facility

Kanban - The strategy of decreasing waste by procuring the inventory you need, only when you are in need of it

TPM - Reduce downtime and enhance total production by using the Total Productive Maintenance process

Kaizen is effective because it is flexible and could be used through the lean tools that are most suitable for your organization.

Identifying Problems

Kaizen begins with the identification of a problem. All people in the organization should recognize that there is an existing problem. Without problems, there is no chance to improve.

A problem in an organization is anything, which results to inconvenience by people involved in the process or by people who are the ultimate end user of your product or services.

The concern here is that the individuals who usually create the problem are not affected by it, and so they may not be sensitive to the problem. In the daily management situations, the natural instinct for employees when faced with a problem is to ignore it or hide it instead of facing it head-on.

This scenario happens because no one wants to be known as the harbinger of problems. But through positive thinking, you can transform every problem into a great opportunity to improve. If there is a problem, there is also the potential to improve. If a problem has been identified, it should be solved. If the problem is solved and the standards have been surpassed, then there should be new standards.

Setting New Standards

Improvement is impossible if there are no standards. The origin of any improvement is to precisely know where you are standing. There should be a specific standard of measurement for each personnel, each tool, and each process. Likewise, there should be a certain stand of measurement for each manager.

Kaizen philosophy requires constant improvement. It is a continuing challenge to the established standards. For Kaizen, the standards are in place to supersede better standards. Kaizen is based on constant revision and upgrade. Only vital elements should be standardized and measured.

Setting standards is an effective way of disseminating the advantage of improvement for the whole organization. Every staff from the CEO to the workers must know the standard and follow it. This is known as discipline, which is the foundation of Kaizen.

The System for Welcoming Suggestions

Among the primary channels of attaining Kaizen and involving all personnel is through the system of welcoming suggestions. Take note that this system does not always require for an instant monetary reward for every suggestion.

This is perceived as a way to boost morale, which can be improved through Kaizen as every person in the organization is mastering the art of solving immediate problems. Many companies in Japan regard the number of suggestions made by every worker as a vital criterion in evaluating the performance of managers.

Meanwhile, the supervisor's manager is encouraged to help them so that they could help the workers to come up with more suggestions. Management is often willing to recognize the efforts of employees. Usually, the number of suggestions is individually posted on a certain are of the workplace in order to drive competition among departments and workers.

A common Japanese workplace has a reserved corner used to publicize company activities. This corner also includes the present level of suggestions as well as the latest achievements of work teams. There are times that tools, which have been improved as a result of the suggestions of workers, are posted on this corner, so other personnel can also adopt the same idea for organizational improvement. The starting point of Kaizen is for the worker to follow a positive attitude towards improvement.

When implemented, every suggestion could lead to the revision of the standard. But since the new standard has been established because of the suggestion coming from the workers, there will be organizational pride in the newly set standard, and so the workers are more willing to follow it.

Meanwhile, if the standard has been only imposed by the management, the personnel may not be that willing to follow it. Therefore, through suggestions, personnel could participate in Kaizen in the workplace and play a crucial role in improving standards. Japanese executives are more confident in implementing suggestions from their employees compared to Western executives.

Managers who are aware of the Kaizen philosophy have more leeway to implement the change if it could improve the organization, specifically in order to:

- Make the work easier
- Eliminate dullness at work
- Eliminate nuisance at work
- Increase safety
- Increase production
- Save time and cost

Process Focused Mindset

Kaizen could lead to a mindset that is focused on the process, because processes should be improved before you can obtain improved results. It is also focused on the people, and it is directed towards the efforts of the people. This is in sharp contrast to the results-oriented mindset of Western culture. In the Japanese workplace, the process is considered just as essential as the expected result. In the American workplace, regardless of how much you have exerted effort, poor results will lead you to poor ratings. The contribution of the individual staff is only valued for its tangible result.

A manager who is focused on the process is more interested in the:

- Use of resources
- Discipline
- Involvement and participation
- Skills development
- Productivity at work
- Communication
- Morale

The Kaizen Manager is more people-oriented and develops a reward system, which is based on the factors above.

Innovation vs. Kaizen

Innovation versus Kaizen can be regarded as the great-leap forward strategy versus the gradual change strategy. Japanese society prefers the gradualist strategy and American companies prefer the great leap forward, which is a strategy epitomized by innovation.

Innovation is highlighted by significant changes in the emergence of scientific and technological breakthroughs or the rise of new production strategies or management processes. On the other hand, Kaizen is non-grand and usually subtle in approach. Its results are often not instantly visible. Innovation is a one-shot approach, while Kaizen is continuous. Moreover, innovation is money and technology-oriented, while kaizen is focused more on people.

Application of Kaizen in the Toyota Production System

One of the many popular stories about the actual implementation of Kaizen in the workplace is the Toyota Production System. Toyota uses Kaizen as one of its central business philosophies. If you have attended business school, you probably heard this popular story about how Kaizen works in Toyota assembly lines.

According to the business legend, Toyota strictly applies Kaizen that any personnel working on the assembly line has the calls to stop the production at any time to resolve a problem, rectify an error, or to recommend to the managers a better way to do things that could improve efficiency or reduce waste.

The popular story goes like this: Auto executives from the United States visited the car plants of Toyota to know how the Japanese workers were able to produce thousands of cars every month with very few errors and very minimal waste.

In contrast, the US plants were hitting the production quota, but there are many errors during the production. An error somewhere along the assembly line like a misaligned steering wheel, badly soldered door, or misplaced bolts — will make it all the way to the last stage of production. The error will only be detected during the quality control, and then the whole car would be disassembled and then reassembled again to rectify the mistake. This process costs more money than if the mistake had been resolved instantly, or if it were avoided in the first place.

The American executives observed the process of Toyota production and were amazed by how effective it is. It is rare in the US for assembly line personnel to cease the whole production line without the approval of a supervisor. The concept of rewarding personnel for instantly correcting errors — even if it wasn't included in their job description — was unheard of in the US; particularly, the motto in the production plants was "never stop the line." The American executives went home and then adapted similar processes. They began rewarding workers who find out better ways to complete work with fewer errors and working with high quality.

As wonderful as this business legend could be, those principles are inherent to Kaizen as a philosophy for achieving better productivity. Once you embrace Kaizen, your goal is to perform better work and not more work (Quality over Quantity). Likewise, it is essential to make time to search for optimizations and improvements.

Take note that the goal of welcoming Kaizen to your life is not only for the sake of changing something. Kaizen can help you be better at work — and so your life.

Most people believe that a company exists only to create profit. However, this is not exactly true. If you were to ask any CEO of a successful corporation why their organization exists, they would tell you that their main purpose is to create value for their constituents. In other words, a business exists to serve the owners, shareholders, and customers.

This value is created by utilizing the available resources and generating outputs that are greater in value than the inputs. This entire process chain cannot be profitable unless the business processes are effective and efficient.

So what does all this have to do with Six Sigma?

You are going to learn how Six Sigma can be used to ensure that the business process generates maximum value at a minimum cost. You will discover the philosophy and principles behind this methodology and some of the benefits incurred when Six Sigma is adopted in an organization.

Defining Six Sigma

Six Sigma can be defined as a system of statistical-based tools, techniques, and methodologies that are designed to eliminate defects and errors in products and services while minimizing process variability. A defect is anything that does not meet the customer's expectations. In simple terms, Six Sigma is a system that ensures that your business produces products and services that are of a *consistently* high quality.

If you take a good look at the business world today, you will realize that every organization worth its salt maintains an online presence. Since we live in the information age, news tends to spread like wildfire.

Therefore, companies have been forced to raise the quality of their goods and services, lest a disgruntled customer grabs their phone and tweets or posts a comment about bad service. Businesses must now guard their reputations and brands by making sure that they consistently provide products and services that meet the customer's needs.

Of all the possible quality systems available, it is Six Sigma that has been accepted as the standard.

But how does this relate to performance?

Sigma is usually represented by the Greek letter, σ, and refers to a measure of variability. If you want to determine the performance of a business, all you have to do is measure the sigma level of the processes it uses. It is important to understand that every process has an average or mean value.

Any deviation from the mean is considered undesirable, but since no product or service is perfect, some allowance needs to be made. This is why the Six Sigma process is divided into six standard deviations from the process average.

We have 1 σ, 2 σ, 3 σ, and so on until the maximum of 6 σ. Most companies usually have processes that perform at 3 or 4 σ, but the goal of every company is to attain a level of 6 σ, which is based on defects per million opportunities. The standard for Six Sigma is 3.4 defects per million opportunities. However, the good news is that even if you do not attain the 6 σ level, any improvements you make between 3 σ to 5 σ will lead to a substantial decrease in costs and an increase in customer satisfaction.

The Philosophy Behind Six Sigma

Six Sigma is a scientific methodology, and as such, you must use scientific techniques when developing your business processes and systems. It is these scientific techniques that the employees will use to improve the value of the products and services. This will ultimately benefit both the customers as well as the shareholders.

As a philosophy, Six Sigma embraces the idea that every individual business process can be measured and then improved. Here is a simplified way of looking at how the Six Sigma philosophy is used in a business:

1. The company identifies one key aspect of its business, maybe a process that is considered to be underperforming.

2. A hypothesis is developed. This hypothesis must be consistent with what is being observed.

3. The company runs some experiments to confirm their observations about the process. As more observations come to light and fresh data is recorded, the hypothesis is adjusted.

4. Statistical methods are used to separate real data from noise.

5. Steps 2, 3, and 4 are repeated until the hypothesis matches the actual results.

If the company continues to follow this system over an extended period of time, it will be able to come up with a theory or model that enables it to easily understand every internal business process as well as its customers. The end result is that instead of making decisions based on hypotheses or guesswork, management will begin to rely on hard data.

Though most organizations think that they operate on real data, the truth is that they do not. Some are simply being run based on traditions. This is why you will hear a manager say, "That's how we have always done things around here, and it works."

Yet the reality is that a company that focuses on using the Six Sigma approach systematically and consistently will improve its performance over time and leave the rest drowning in its wake.

Principles of Six Sigma

To guarantee success when using Six Sigma, you have to learn these five key principles:

1. Customer focus

Before you improve the quality of a product or service, you must first define the word "quality." The best person to define what quality means is not you but your customer. Therefore, focus on the feedback from your customers and adjust your processes accordingly.

2. Identification of causes of variations

When you are dealing with a business process, you need to understand that variation is your enemy. Your products must be consistent in terms of quality. If customers cannot trust that what they buy today will be of the same quality tomorrow, they will flee to your business rivals.

The first step in identifying the root cause of a problem is to understand how the process works. Not how it is expected to work but how it actually works. To achieve this, you need to:

- Identify the kind of data you need to collect
- Clearly define why that data must be collected
- Clarify the information that the data will reveal
- Communicate the terms clearly

- Make sure that the measurements taken are precise and repeatable

- Develop a standardized system of data collection

Data collection is generally done by interviewing various stakeholders, observing the process, and asking the right questions. Examples of questions that you should ask include:

- What can we do to make your job easier?

- Why is this process done this way?

- Are there any tasks that you perform that seem unnecessary?

Once you have collected the data, use the information to find the underlying cause of variation.

3. Elimination of variation

The next step after identifying the root cause of variation is to eliminate the variation. To achieve this, you have to make adjustments to your business process. These adjustments include eliminating any steps that don't add value to your customer. The end result will be the elimination of defects and minimization of wastage.

4. Teamwork

To ensure that Six Sigma is implemented properly, you need to have a diverse team that is committed to incorporating the Six Sigma methodologies. When a team has members with diverse skills, they will be able to detect variations in the different areas of the business process.

The team must be highly skilled and trained in the use of Six Sigma tools and techniques. There are five levels of Six Sigma certification. These are Master Black Belt, Black Belt, Green Belt, Yellow Belt, and White Belt. The highest certification is the master black belt.

5. Flexibility and thoroughness

To successfully implement Six Sigma in an organization, the system must be willing to accept change. Management and employees must be flexible in their thinking so that they see the benefits of the changes being implemented. This means that the employees must be clearly informed about how the changes will affect their work.

If they are consulted early, Six Sigma will be readily accepted. The implementation itself must not be so complicated that people would rather stick working with a flawed process than move to the new one. You also need to make sure that you thoroughly understand every area of the process.

Chapter 1: What is Lean Analytics, A General Overview and a Little Bit of History about It

What is Lean?

The lean concept is nothing new. It has roots tracing back to the 1950s when Toyota shifted its focus to optimizing product flow through the entire production process. The company introduced machines that both met the needs of the volume and demand as well as monitoring machines that would ensure the proper quality of each product. As a result, Toyota was able to reduce the cost of production and increase quality and output (A Brief History of Lean, ND).

The key principles of lean, as presented in James P. Womack and Daniel T. Jones book, Lean Thinking, are described as:

1.**Value:** Having specific values that are desired by customers.

2.**Value stream:** Being able to identify the unique value stream for a product as it relates to customer value and eliminate unnecessary steps.

3. **Flow and Pull:** Ensuring the product flows smoothly through additional value-added steps.

4.**Perfection:** Constantly focusing on reducing the amount of time, steps, and information needed to supply the product to the customer.

These steps should continuously flow from one to another, starting with identifying the value, mapping out the value stream, then creating the product flow, determining the pull, working toward perfection, and then returning to identifying the value.

The lean Start-up offers a similar approach with its feedback cycle or loop:

•**Build**

•**Measure**

•**Learn**

Here, you create a plan that defines what needs to be tested and what you think the results will be. Then, you determine how you will measure those results and collect data. Next, you build a product that is small so that you can test out your thinking. Once the experiment is conducted, you measure the results you gathered. How do these results compare to your initial thinking? Finally, you learn what to do next and repeat the process when necessary.

Keep the following concepts in mind:

1. Know the type of business you are.

2. Know where your business is at.

3. Track the One Metric That Matters and optimize this metric.

4. Repeat the process.

The fundamentals of Lean Analytics

Lean, Lean Start-up, and Lean Analytics all have different methods, cycles, and stages. While they rely on different approaches, they do have a few fundamentals in common. The three fundamental issues that should guide businesses are the purpose, process, and people.

Purpose: Focuses on the problem they will solve for the customer.

Process: Assesses the value streams to ensure it is valuable, capable, and available as well as adequate and flexible. A value stream is similar to a map that tracks all the business actions taken that in some way, create a product or value for the customer. These should help link each step of the process with the flow, pull, and leveling.

People: Who will be responsible for evaluating the purpose and process? How can engagement be increased and encouraged among those involved in each value stream?

The ultimate goals and focus of implementing the lean Start-up method and lean analytics are to minimize waste, continuously improve upon the business idea, and always have the big picture in mind.

The lean Start-up runs off a simple cycle of collecting and measuring data to help improve on the key fundamentals. Lean analytics is the process of not just gathering data but also knowing when and how to analyze that data to understand where your business is and where your business is headed.

Lean analytics takes a more scientific approach to develop a product or business idea, where you first address the problem that needs to be solved, then go about trying to find a solution and ultimately create experiments that will allow you to test and measure the results.

Waste and the lean system

The goal of the lean process is to eliminate waste completely. Waste, according to the lean Start-up method, is anything that doesn't lead to validated learning. Instead of keeping all things separated for the development and growth of a business, things are instead able to flow together in a more unified way. This results in reducing the need for space, time, and effort, which, in return, lowers the cost.

Waste can include:

•The overproduction of a product or service, when something is produced before it can move onto the next process. This can occur when the product is made too quickly or in excess.

•Inventory waste refers to producing large quantities of a product and then having to manage this surplus of inventory.

•Motion waste occurs when there is a poor design in the work environment.

•Transportation waste refers to the poor system of moving parts of the business from one place to another.

•Over-processing waste results from confusion with the customer when items or materials are overly complicated.

•Defect waste includes poor labeling, inadequate information, poorly written instructions, and other details that do not fully disclose or explain parts of the product or service.

- Waiting waste can be incurred from the customer who either has to wait in line for customer service or has to wait for an extended period of time before they receive their product. It also relates to the production side of having to wait for one team or member to complete their job before another can begin.

- Underutilizing staff waste can occur when there is a lack of communication or a flaw in the managing system.

Waste can hinder and significantly slow done project development. When you find a system, like lean analytics, to greatly minimize waste, you speed up the Start-up process. Since lean analytics stresses the importance of focus on one metric at a time, you do not get distracted or waste time in areas that won't move the process forward.

How can lean help define and improve the value system?

This lean approach helps you identify where your system has flaws and stops flowing. A good value system is one that is predictable and moves through each stage or phase without halting. A consistently good flow is one that moves faster and offers customers greater reliability.

The value stream identifies either how your vision or the value you have ultimately reaches the customer. When directing this path, a useful tool to guide you is a Kanban board. Through a Kanban board, you will be able to map the direct steps of the value stream visually. Kanban boards help divide the value stream into the following three components or columns:

- **Requested**
- **In progress**
- **Done**

These boards are an easy way to visually see where workflow is becoming backed up and what needs your attention to continue to progress forward. These boards are ideal to use with lean analytics because they help you stay focused on the tasks that need to be done. When it comes to the value stream, these boards provide you with clear visuals on where the stream is being stopped or slowed down.

Lean analytics puts the customer as the first step in understanding how the stream will flow. It is the customer which in the end, will be the driving force and decide on the flow of the stream.

In 2004, he and a few aspiring individuals decided to come up with an instant messaging market. Instant messaging was still fairly new at the time, so the potential for creating a successful business in this industry seemed straightforward and easy to navigate.

Immediately, the group began to create a product that would incorporate all instant messaging services around at the time and allow users to choose avatars to chat with one another online no matter what messaging network they or the other person was on. The creators thought that not having to learn a new platform or switch to a new one would greatly appeal to users. They also believed the product would take off virally and quickly gain users from those on every messaging platform.

For six months, the group worked on a website that integrated all messaging platforms into one place. When they finally launched it, they eagerly waited to see users quickly jump on board. But nothing happened. They could have easily blamed this on the fact that this was the first edition and it still had a few bugs that needed to be fixed, but the fact was that no one had even bothered trying the product.

While the group still worked on improving the site on a nearly daily basis, only a few users had bothered to sign up. To try to understand what they needed to do to improve the site, they asked people to come in to try out the site. After allowing a small group of individuals to try out the program, they realized where it was flawed.

They had believed that creating a completely new messaging service would deter users because they assumed users wouldn't want to have to learn a new program. It turns out, that's actually what users wanted. The users enjoyed being able to create and customize the avatars, but they didn't want to have to go through the process of integrating it on the eight or more messaging channels they were already on.

What Ries realized through this experience was that they wasted a lot of time and energy creating a product they thought users would love, but never once did they think to ask the users if they would use it. Had they started the process by first finding out the customer's needs, they would have been able to avoid wasting their effort on a product with little chance of success and instead could have immediately begun coming up with a product that fits what users were looking for.

This is how the Lean Start-up idea began. Ries found a major flaw in the project development system. Instead of focusing on first creating the perfect product, entrepreneurs have first to begin to understand what that perfect product is.

Getting started

Having the right tools to assist you can be of great help. Lean Canvas is a tool that helps you organize your business idea from beginning to end and helps you identify key metrics to test through each stage.

The lean canvas approach allows you to identify key factors from the beginning to the end of your project development or business ideas. This can be done in a few simple steps:

Step 1: The Problem

The problem is the whole reason you are beginning your business. There is a problem that the customer needs to have resolved. By now, you already know what the problem is, but do you also know three existing solutions for that same problem? List your problem first, and below that, list the three existing solutions you know your customer already uses.

Step 2: The Solution

After the problem, you will list the solution you have come up with. For every problem that you write in the first step, you should have a relating solution in this step. You also want to write down the top three key features or functions your solution offers that further help solve the problem.

Step 3: The Key Metrics

As you just learned, metrics are the most important factors in your business. In this step, you want to list the key metrics that are both relevant to where you are in your business and that you think you will need to keep track of in order to stay focused on the solution.

Step 4: The Customer Segments

After all the research you have done, this should be an easy section to fill in. Here you want to identify your intended audience. You should be able to easily come up with three to five key customer characteristics of your early adopters. These individuals are the ones who are eagerly awaiting the launch of your product or service. Understand what makes them so interested in what you have to offer.

Step 5: The Unfair Advantage

You likely have a few things that put you ahead of your competitors. These unfair advantages that you possess shouldn't be easy to copy or acquire. They can include things like inside information or an in-depth understanding of the problem, being an expert in the industry, having a large supporting network or community, having the ultimate team, and having a highly respected reputation. These are things that you will want to use to your advantage through the process.

Step 6: The Channels

What are all the ways you can contact or reach customers? List all the social media sites, people, networks, and touchpoints. Touchpoints are where customers will most likely encounter your product or brand. There are three main time frames that you will want to break this up into:

- **Before purchase**
- **During purchase**
- **After purchase**

In each time frame, you want to list the top three channels of how you will contact or connect with your customers.

Step 7: The Unique Value Proposition

This statement answers the "how" and the "why" of your business model. This should be a clear statement that highlights how your business stands out and places value above any other alternative. You want to write your unique value proposition and look at it as a high-level concept as it will be the statement that helps show customers what they should expect from your business.

Step 8: Cost Structure

Here you will list all the costs you can think of that will occur when doing business. Look at each step you have completed so far and consider the cost that can be attached to each step. One of the main reasons start-ups fail is because they do not properly plan for how much it will cost for them to launch and start their business. Using lean analysis can greatly reduce these costs, but you still want to address all. Some cost can include:

- **Customer acquisition**
- **Distribution**
- **Product development**
- **Services to launch**
- **Connecting with customers**
- **Branding**
- **Researching your market**
- **Marketing**

Step 9: The Revenue Streams

What are your sources of income? How will this source of income continue to keep your business running? Some of the most common revenue streams include?

•**Asset sales**

•**Usage fees**

•**Subscription fees**

•**Delivery and installation fees**

•**Advertising**

Once you have completed all the steps, you will have a simplified business plan that will get you started and help you stay focused on your business idea.

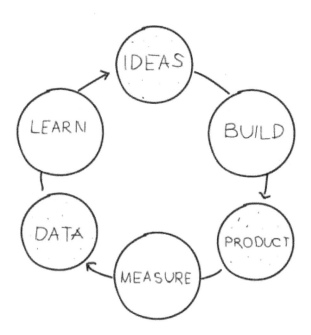

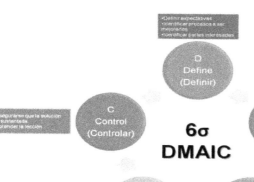

8 Wastes: Lean Six Sigma

Inventory	Talent	Waiting	Motion
Excess products and materials not being processed.	Underutilizing people's talents, skills, & knowledge.	Wasted time waiting for the next step in a process.	Unnecessary movements by people (e.g., walking).

Defects	Transportation	Overprocessing	Overproduction
Efforts caused by rework, scrap, and incorrect information	Unnecessary movements of products & materials.	More work or higher quality than is required by the customer.	Production that is more than needed or before it is needed.

Chapter 2: Lean Thinking

Best Practices in Lean Thinking

Like every project management methodology, lean thinking comes with its best practices. As mentioned before, the purpose of this book is not necessarily to introduce you to general Lean thinking, but to help you understand Lean Six Sigma and how it can be applied to your team and projects. However, you can't understand Lean Six Sigma and the way it functions before you understand Lean thinking - so following, we will present you with some of the basic best practices connected to Lean project management and thinking.

1. Focus on demonstration, rather than explanations. This is one of the fundamental rules of Lean thinking and, in general, a rule that can be applied to many other project management methods.

2. Encourage Kaizen activities. Workshops, team quality circles, suggestions you receive from the team members, and general exercises focusing on continuous improvement - they all help the team grow in the spirit of Kaizen, Lean thinking, and, in general, in the spirit of constant improvement. Do keep in mind that this kind of

activities should be thoroughly planned and that they should not be left at the bottom of your priority list: they can truly make all the difference in the world in terms of how your team perceives Lean project management and thinking.

3. Kanban. If Lean thinking is the philosophy, Kanban is the foundational practice (also used at Toyota originally, and these days, all around the world in multiple industries and contexts). Kanban helps managers determine what is necessary to be done *now* so that the customer is provided with the best value in an ongoing way. For instance, if you are a writer and you have to write books, create blog posts, and answer emails, you can use Kanban to determine which of these tasks are more *current* in terms of customer satisfaction. Is it mandatory that you finish another book by the end of the month when the old one is still selling, or would you much rather focus on PR and promotion activities, such as writing on your Social Media channels or replying to emails?

4. Autonomation. In a world that seems to automate everything, there are still many instances when it is absolutely crucial that humans are involved in the production process. Autonomation enables human help to be asked for only when the machine "feels" it has done

something wrong. This way, team members can focus on other tasks, machines can work properly, production can be kept at high levels, and customers can be kept happy.

5. SMED (Single Minute Exchange of Die). This Lean thinking concept is all about flexibility and training team members to change their tools in under 10 minutes. In other words, SMED encourages team members to quickly switch between activities.

6. Standardizing the work. While Lean thinking holds flexibility sacred at its core, it is also very important to note that standardizing the work is crucial when it comes to this project management approach. The reason standardized work is so important is because it creates a smooth flow - one team members can follow and one that can be easily predicted, so that you can constantly plan ahead new inventories, according to the current demands of the client.

Lean Thinking to become Lean Enterprise

Lean is a journey: a never-ending pursuit for perfection. Lean Thinking improves the processes by the reduction of the wastes. A philosophy which focuses on the removal of the unnecessary steps or procedures involved in the process. It is customer-oriented and the steps which need to be eliminated are determined based on the customer's perspective. It is focused on the business transactions, but this thinking can be easily implemented in our daily lives to get the maximum amount of efficiency out of it.

Identify value stream

Value Stream is the flow of all the processes which include all the steps from the initial design, development, launch, and order delivery of the products and services. Although 100% perfection can not be achieved yet, lean methodology reduces the wastes to the minimum value and maximum value added processes. According to Lean Thinking there should be a constant communication between the customer, producer, and the management to reduce wastage.

Value Stream mapping

Value stream mapping is an amazing tool that helps to identify major non-value add activities (wastes), which must be removed from the process to make it lean.

What do we mean by a value stream map?

A value stream map is a graphical representation of all the activities which constitute any process under consideration. The activities represented in the value stream map can be essential activities, wastes or non-value add business activities. It contains a lot of information regarding the process under consideration and is extremely helpful in understanding the flow of the procedure.

What do we get by drawing a value stream map?

When a value stream map is constructed, understanding of the mechanism of the flow of activities and their significance becomes clear to the management and anyone studying the value stream map. It also helps to identify nonessential steps in the process that must be eliminated from the process to make it lean.

Tips for developing a value stream map

The value stream map is a simple tool for making the business lean. If applied efficiently, it can result in great value generation with minimal investment of time, mental capabilities, and physical efforts.

- **Use Sticky notes**

Sticky notes are fun to work with, but that is not their main appeal or attractive feature. You can comfortably make changes in them and you can color-code them as well. For instance, you can designate green colored sticky notes only to be used for essential activities, red sticky notes for wastes, and grey colored sticky notes for non-value add business activities. This way, it becomes easy to identify the different types of activities when the value stream map is studied.

- **Make sure that your workstation is spacious**

When developing a value stream map, things can become very messy very fast. If you are working on a value stream map in a congested space, it will become very difficult to avoid cluttering up different things. The more spacious the workstation is, the easier it will be to manage it. It would be much preferable if you work on a big whiteboard or a giant desk when you are developing a value stream map.

- **Don't develop the value stream map all alone**

It is best to develop the value stream map with a team of professionals who are personally involved in the process. It eliminates or reduces the possibility of overlooking a step or classifying an essential activity as a waste or vice versa. It also allows you to have an eagle's view of all the steps involved in the process and find the loopholes in the process.

Benefits of value stream mapping

- Highlighted dependencies
- Identify opportunities
- Understanding of the highly complex systems

- Synchronized and prioritized continuous improvement activities

Types of value stream maps

- Production: raw material to the customer
- Design: design to concept launch
- Administrative: order-taking to delivery

States of value stream :

There are two states of the value stream, which are as following:

Current state:

The existing conditions in the value stream is called current stream.

Steps for making a Current state map

- Determine the type of map using the flow chart. At first be very general and add uniform details as you go along. Pay particular attention to the critical paths. Add the elements such as inspection and test, also include the waste for its productivity is of equal importance.
- Add inventory, transportation, vendor facilities and customers endpoints.
- Attach functional groups and information flows.

- Develop and attach data to all elements such as lead times, setup time, and process times.

Future state:

The Future state reflects the future vision of the value stream.

Steps for making a Future state Map and Work plan

- Use the current state map as base line.
- Using the 7 Waste type definitions and analyze one at a time to see which element contains waste and attach a measurement of the waste.
- Estimate the use of resources required to accomplish the changes. Calculate the human resources requirements and don't over-estimate the available resources.
- Redraw your map consistent with your change selection.
- Make a detailed work plan of who, what, when and how the processes and activities would occur. Processes should be reviewed regularly, if the planned should be changed, it should be discussed in advance.
- At the end of the plan, adjust the map to reflect the changes. This will now be the current map. Decide

whether to go for another cycle or to change the map subject.

1. Make Value-Creating steps flow

Flow includes all the steps through which we go along through the value stream with no wastage or faults to achieve the desired goal. Flow reduces the waste that creates hindrance which stops the value chain to advance forward. An efficient value stream should not hamper the manufacturing process. All the activities from design to launch of the products or the services should be synchronized, which will help in the reduction of waste and will improve the efficiency. Customer happiness is the most important to make value flow.

2. Pulls Customers towards product or services from value stream

Traditional business systems are such that they produce products in large bulk, hence the quality of the product falls. The bulk of the product is then stored away hoping it would find market. This is known as "push system". While the Lean Thinking promotes the "push system". According to this system the manufacturers do not make a large amount of product and store it, rather they make goods which the customers demand. The value stream pulls the customers towards the products and services. There the manufacturers would not make anything unless it is demanded by the customers. If a company is following Lean Thinking then it should perform quick actions and a lot of flexibility. As a result, the cycle time required to plan, design, and deliver becomes very short. The biggest advantage of the pull system is that no values activities can be minimized.

3. Perfection

To attain perfection is the main goal of lean thinking. This is because continuous improvement is required to sustain the process. And to sustain a process it is important to remove the reasons which are behind the low quality of the products and services. Lean masters who are individuals from various teams with a common goal. The goal to achieve efficient results. The results benefit either the organization or the customers. The most effective way to achieve perfection is by the collective effort of engineering, supplier associations and value stream mapping between customers and suppliers. It is significant that the lean principles should be followed to reduce waste, deliver quality goods to the customers and gain maximum profit. The organization and the customers should work together to achieve the desired goals and the visible efforts to reduce waste and improve efficiency. Lean Thinking can be applied with the help of committed leadership, a persuasive change agent, and well-informed employees and suppliers.

The same idea that we can apply to process improvement can also be applied to product development and vice versa. So the idea of optionality can be applied to both process improvement and well as product development. You can assign a bunch of teams, they all try different things and see what works and what does not and then gets adopted by the teams.

Making business Agile

Start with the business objective. If you don't have the business objective, you don't know how to do things. It is less important how the teams work but it is important how the teams work together. Figure out organization optimization and to bring value to the customer.

Make sure that at the enterprise level, whatever level it is, you decide where the teams can engage with the executives and transform the business. It needs a continuous improvement process. Rather than writing done hundreds of stories, sit down and plan out strategically what you are trying to accomplish as an organization, then it will start to add value.

If the teams know that there is something on the list that they need help with, then they will be able to access the resources that are available in the organization and to help them get done. If there is nothing in the priority list, then the teams keep doing what they are doing at the team level. It is a nice combination of tops and bottoms up. It gives a strategic direction going and gives empowerment to the teams.

In order to achieve a milestone, the teams and the programming staff gather together and talk about the steps to take to achieve the goal, what should be done and what they learned, the data they collected and then draw out the rough draft of what they thought the objectives would be. Rough draft contains the opinions at the time which need to go through staff meetings, organizations and forms to see if they are not missing anything. The objectives should have everything that an organization is aiming to deliver but should also include a continuous improvement process.

Hypothesis-driven delivery

The concept of hypothesis-driven delivery is that we believe that building some features for the customers with achieve their desired outcome, hence we make a prototype and receive the outcomes from the experimentation that whether out hypothesis is correct or not. Then we feed these things in our target conditions along with process improvement goals, we have product improvement goals which we have defined through customer outcomes.

The key things that huge companies like Google and Amazon do, they run experiments on processes in production through tests, gather data and often build whole features. The reason they do that is in general the data they gather shows that only ⅓ of their ideas were successful.

Changing the culture (high trust culture)

The heartbeat of what's happening within the team, they outline the aspiration of what's happening within the team, they outline the aspirational vision of what they are trying to achieve, what are the goals, what are the hypothesis they are trying to test and get them in front of customers. What experiments they are going to run this starts to map or design the picture of how their product or story fix together but before that they collect the data and define the measure of success to validate how they can run those experiments and what are they going to learn.

Aspirational vision
Customer hypothesis
DESIGN experiments
Story mapping
Data/feedback to validate

The introduction of such processes and practices in an organization changes it because of the changes in the way people work. In order to change culture people's behavior should be changed. In order to change people's behavior, change the system of work in which they operate within. Make a system in which they are happy and they can start to operate to achieve the goals.

Building the Lean culture is the key to innovation. Creativity must flow from everywhere in the organization. Whether it is an intern or a CTO all the ideas must be exposed to objective testing, experimentation and preferably a test that exposes the idea to real customers. Everyone must be able to experiment, learn, and iterate.

CONCLUSION

Sometime in the 1980s, one of the greatest movies about martial arts was created. *Karate Kid* came like a storm - and even well into the 2000s, televisions still broadcast the movie, with nearly the same periodic recurrency as *Home Alone*.

There is a very good reason people loved that movie (as bad as it may be): it was endearing, it was about martial arts, and it gave everyone hope.

Sometime in the 1980s, one of the greatest project management methods ever created was born. *Lean Six Sigma* came as a result between Lean and Six Sigma - and it took the world by such amazing power that even today, people keep perfecting the theory, people keep learning the rules, and people still use it to save companies thousands of dollars after thousands of dollars.

Aside from the decade they were born in, and aside from the fact that they both took the world by storm, *Karate Kid* and Lean Six Sigma have one more (very important!) thing in common: their reliance on martial arts philosophy.

Sure, *Karate Kid* is but a sketch of what it actually takes to win a martial arts competition - but even so, the endearing messages and the best moments of the entire movie are the ones connected not to actual martial arts theory, but to the more romantic aspects of fighting for your title.

In Lean Six Sigma, just like in martial arts, you start low. Like Daniel-San sweeping the windows of the car in imperfect hand collaboration, you will first find that handling all the aspects of Lean Six Sigma feels a bit overwhelming - and, at times, you might even feel as if you are writing with your left hand.

The similarities don't stop here. Just like in Lean Six Sigma, Daniel had a teacher - a mentor to show him the intricacies of martial arts. And just like companies using Lean Six Sigma, Daniel-San's teachings were all about balance and routine processes that help him find his inner core of strength.

Given the fact that even the titles in Lean Six Sigma are inspired by martial arts, the comparison between this methodology and *Karate Kid* is not far-fetched in any way.

In fact, it can be fairly assumed that if you are ever in doubt about Lean Six Sigma theory and philosophy, you can simply think of a martial arts master and think of what they would do in your given situation.

Chances are that you will find an answer that is at least close to what Lean Six Sigma would propose.

Lean Six Sigma is fascinating to people for a billion reasons - and its martial arts-based nature is one of the reasons that attract curious minds towards this project management approach. It makes sense that everyone wants to become a Bruce Lee of the project management world, right?

Lean Six Sigma is a truly amazing method to employ - as long as it suits your company, of course. As it has been shown in this book, not every business and every project is meant to be applied to a Lean Six Sigma approach. In some cases, this theory is just not suitable, and it would not bring anything valuable to the table - so if you find yourself in this situation, keep the information learned in this book for "later." The chances are that you will, sooner or later, use it to fix some sort of process error in your company.

Not only is Lean Six Sigma pretty awesome from the point of view of the symbolism it employs in its naming and methods, but it is a very modern methodology as well. Back in the 1980s, people might not have cared as much about the waste reduction from an environmental point of view - but these days, this is the main buzzword you hear everywhere. And Lean Six Sigma was *there* long before "it was cool!".

We truly hope the book at hand has opened your appetite for Lean Six Sigma and everything it comes with.

While this is not even by far everything there is to Lean Six Sigma (we could talk about it for another 100 books), we hope the book at hand has helped you understand the high-level theory around this project management and problem-solving method.

We did not aim to uncover all of the Sigma secrets (or the Lean ones, for that matter). We aimed to give you a taste of just how useful, just how interesting, and just how awesome this entire framework can be. Hopefully, you have enjoyed your journey with us.

This is the second time we are using the word "journey" - and it is a very carefully chosen one, mind you.

The first one is related to the fact that once you embrace Lean Six Sigma, you will fall in love with its intricacies, with its symbolism, with the way it can actually help businesses grow bigger and healthier in so many respects.

The second one is related to the fact that Lean Six Sigma is all about continuous improvement - and what advocate would you be if you decided one day that you cannot or simply don't want to proceed further with your improvement in the art of Lean Six Sigma?

Last, but not least, the third important reason that makes Lean Six Sigma a life-long affair is the fact that it will keep surprising you, every time you use it. Sure, the theory might not seem like much when you look at it from afar - but when you see the kind of results Lean Six Sigma can bring with it, you cannot but feel really excited and productive!

Lean Six Sigma is not about empty promises of the kind you see on teleshopping advertorials. It's not a one size fits all recipe for success. And it is most definitely not a scam.

Lean Six Sigma is a way of thinking and a way of seeing life itself. When you filter actions through processes and learn to get to the root cause of things, you will be more tolerant, you will understand people better, you will have more empathy, and you will know how to treat even the more delicate situations in a way that doesn't upset anyone.

Lean Six Sigma is a method, a strategy provider, a system. Its roots may be based at Motorola and Toyota - but the system it creates is more than suitable everywhere around the world, for businesses in multiple areas of activity and of many different natures.

Six Sigma speaks internationally. It helps people from all over the world. It pushes businesses forward and, maybe more importantly than anything else, it pushes *people* forward, helping them be better, act better, grow better.

The main goal of the book at hand was not to scare you off with the myriad of information available on the topic of Lean Six Sigma.

On the contrary, actually. As mentioned above, our main goal was that of stirring your interest in this methodology and helping you understand its basics - precisely because we know that diving head-first into the more advanced techniques would feel downright scared.

The book at hand was meant to open the world of Lean Six Sigma to you and help you see that, no matter who you are and what you work, you can always pick up this theory and embrace it from the comfortable sands of a Greek island or from the comfort of your team.

Hopefully, we have provided you with the key to a new world: one where you don't have to constantly run guesswork operations on what is going wrong in a company. One where you don't want to have the responsibility of what would normally be ten other roles in a company. One where you can find actual solutions to your problems and stop "patching" them as if they are scratches on the knee.

More than anything, we genuinely hope the book here has answered your questions on what Lean Six Sigma is, how it functions, and why it can be of the utmost importance in your future.

If you are the owner of a startup, you will find this method for process improvement to be really useful.

If you are a project manager in a large company, you will definitely find Lean Six Sigma to be beneficial too!

If you are a healthcare worker, you will find that Lean Six Sigma can help you reduce the waste in your hospital so that you can focus on what you know best: saving lives.

Lean Six Sigma can be just the framework you are looking for, no matter who you may be and where you may work.

Therefore, we truly hope this book has helped you shed some light on the steps you should be following from here on, on the main philosophy behind Lean Six Sigma, and on the main techniques of its employees.

Last, but definitely not least, we hope this book was *fun* for you - because what would a learning process be without a bit of entertainment in it? Your future is about to become better because you will implement Lean Six Sigma - so what is there not to be happy about?

We wish you a cheerful, successful Lean Six Sigma road ahead of you. There might be bumps along the way, but trust us when we say that *it is all worth it*!

Scrum can help teams deliver great products on time if the team members, the Scrum master, and the product owner already have the right skills and abilities to create the product. Scrum is not a magical set of rules that any organization could just follow like a cookbook recipe and expect instant results. What I've given you is a basic understanding on the essentials of Scrum and how to use it to tap into the skills of the team members, Scrum master, and product owner, turning those into powerful leverages in creating innovative products on time.

Scrum is flexible in a sense that after several projects, it can morph into a completely different framework, perhaps with more effective tools, artifacts, and roles. Nevertheless, Scrum has no marked finish line. There is no end goal, that means you can stop learning.

Being good in the implementation of Scrum is never the end goal of companies. In the same way that studying is not the end goal of students, rather, to learn more effectively, learning to be more proficient with Scrum means you'll be able to help your company reach its goal better.

Some methodologies actually have end states, which is why they have certain levels to reach. Scrum, however, does not make an assumption that there is a state wherein you can no longer transcend the current state. It assumes that you will always find new and better ways of achieving goals. After all, this is the real world, where the best does not stay the best for too long.

No Such Thing as a Perfect Start

I've probably mentioned this quite a few times in this book one way or another, but that's because a lot of people still can't quite get this concept right. A lot of people try to implement Scrum, only to delay the start of implementation because they can't seem to perfect the concepts of Scrum. Ironically, this is exactly what Scrum is against: waiting for the perfect moment before you begin.

We live in a non-ideal world. Ideal conditions only exist in abstract mathematical notations that only a few of us would live to see. Scrum allows changes to happen because of the fact that people who try to implement Scrum will inevitably make mistakes before, during, and after the development process. A team's concept of perfection before product development will inevitably be different from its concept of perfection during and after.

If you're worried that you don't have everything perfectly planned, you should stop worrying! Perfection means no longer being able to learn new things, and Scrum emphasizes the need to continuously learn and grow. In most cases, the first few sprints may be somewhat disappointing or even downright ugly, but that's all right. The important thing is that the succeeding sprints start to become better than the previous ones, and in most cases, they really will.

Get started! By starting as soon as you can, you give your company a lot of time to grow.

The first few sprints won't be perfect and neither would the sprints in the distant future, but no Scrum implementation is ever problem-proof. All companies have problems implementing Scrum. Remember that since Scrum helps companies discover hidden kinks and bottlenecks, the companies that find these problems may associate difficulties from the problems to the Scrum framework itself. This misconception is understandable because a lot of methodologies sometimes make work seem a lot easier than it is by hiding potential problems, only to let them pop-up somewhere near the end.

Scrum is a bit more thorough, letting the teams see potential problems ahead in the beginning, middle, and end of the production. The thing is, though, Scrum doesn't tell you how to solve those problems. It can only tell you so much; the Scrum master, product owner, and team member all have to work together to find a solution.

I've said before that one can change some aspects of Scrum should he find more effective solutions, but beware that a lot of people tend to change Scrum, thinking that it'd lead to more efficient operations only to find out that they were slowly reverting back to their old methodologies. If there are dysfunctional people in the organization, the introduction of such a powerful framework that exposes bottlenecks, kinks, and other problems may make them rebellious to the idea of its implementation.

There will be a lot of impediments, especially in problematic organizations, before Scrum could be implemented. It takes patience, consistency, and diligence to properly cement the foundations of Scrum into a weakly managed organization. A lot of people, even the ones with good intentions, will rebel. People naturally resist change, especially ones that force them to change their way of thinking. Help the people involved by easing them into the principles of Scrum and give them a concrete view of the goals you're trying to achieve through this change in methodology and framework. The more people understand Scrum, the less they'll resist its implementation. The less people who resist new implementations, the better your company will be at implementing Scrum.

I don't claim that this book has all the answers you'll ever need. Far from that, I encourage you to keep learning and keep asking questions. Keep challenging existing ideologies in a healthy manner. I hope that this book has given you a lot of insights and ideas about the Scrum framework to help you in your journey in creating and delivering great and innovative software. Good luck and may your visions for your company and your teams come true!

Thanks for making it through to the end of *Kanban*. Let's hope it was informative and able to provide you with all the tools you need to achieve your goals.

Your next step is to observe and plan your transformation. Stop wondering how you can become more lean, agile, and efficient. You just read all about it! Now is the time for action. Now is the time to prepare your Kanban board and visual system to make your life easier and your team happier. Now is the time to lower costs and increase production using a simple and effective method.

While you are planning, get the buy-in from your team, company, stakeholders, and even your customers. Sell them on the benefits of adopting a Kanban system, and stay close to the process, refining as needed, so it is the most efficient system for your business. Remember, the goal of this is to assist your team members in working alongside one another efficiently while also benefiting your company. Keeping your eye on this goal during each decision you make will help with all the changes and challenges.

A Kanban methodology applies to a variety of situations, despite rumors it is "outdated." As it is with new technology, do not jump onto the glossy "bandwagon." Determine the unique needs of your organization and create a way to make this basic system work for you.. The more and more you use boards, lists, and cards, the better your team will get at running an effective Kanban project and process. As they continue to feel empowered and successful, imagine the positive atmosphere and engaged work environment you will have! Success will come to you in a variety of forms thanks to you implementing this methodology in your company. Congratulations on taking this step towards a more productive future for your company!

Inefficiency affects organizations of every type and size. Even in the most positive economic climate, companies benefit from decreasing defects and avoiding wastes. When times are tougher, enhancing efficiency can often mean the difference between keeping the business's doors open and a foldup.

Initiating Kaizen-based continual business process improvement campaigns keep companies on a growth curve.

Kaizen or 'Make Better' requires a mindset shift from an 'okay process' to 'continually better process'. The improvement is made not as a chunk, but as incremental changes accumulated over time. The changes are rapidly implemented, hence Kaizen is also called 'instant revolution'.

Kaizen may be applied to manufacturing as well as service-based organizations. Implementing Kaizen awards competitive edge & consistent growth to a business. Personal Kaizen further improves employee's quality of life both at personal and professional fronts.

Now that you have come to the end of the book, I hope that you appreciate everything that Six Sigma stands for. Over the years, there have been a lot of myths and confusion regarding this particular methodology. However, one thing should be clear by now.

Six Sigma is extremely useful for any organization that wants to improve quality, reduce costs, and enhance the speed of delivery of goods and services.

You have learned the most important tools and processes that are used in Six Sigma implementation. Keep them in mind as you move on to the next phase of the journey – which should be implementing and deploying a Six Sigma project.

Remember to follow the right procedure when trying to identify a solution to defects in your business process. Use the DMAIC stages to guide you every step of the way.

Improve the organization and control over your line, process, area, department, shift or full factory starts with being clear about what you are trying to achieve.

You need good skills around you – enough strength in depth to deal with whatever is thrown at you. A team made up of skilled, experienced people who know what is expected and what to do is the first step in the journey.

Once the team has enough key skills, it will have the capacity to deal with the day to day issues and challenges that occur in factories whilst being able to take on adopting new practices such as 5S.

5S will clear the decks of clutter, make the essentials like tools easier and quicker to locate and will give the area an organized and controlled appearance and "feel".

Keeping on top of regular red-tagging, sorts and sweeps will allow you to maintain this level of organization. You can't do it only once! It must be done until it is so habit-forming that the team do it without even thinking about it. This can take some time but it will happen if you stick with it.

Visual lean techniques like shadow and line marking will reinforce the look and feel of a controlled organized workplace.

SIC and SPC are further tools to bring in place to allow the team to quickly see and understand where they are against where they need to be.

Throughout keep using a DMAIC approach to keep on track, keep communicating to your team and everyone involved. This will create a positive "feedback" loop by allowing people to see the improvements flowing through. Seeing improvements being delivered creates a feeling of positivity and spurs teams on to achieve more.

Keep moving forward, don't quit even when things don't happen as well or as fast as you want them to, and before you know it you'll be seen as a someone who can deliver improvements and change within your business.

CPSIA information can be obtained
at www.ICGtesting.com
Printed in the USA
BVHW042222130421
604819BV00009BA/1231